Interplay of Scientific Ideas

By

Miguel A. Sanchez-Rey

Candidacy and the Grand Unification Scheme

Miguel A. Sanchez-Rey

Abstract

Does any unification candidate necessitate the grand unification scheme?

May 24th, 2016

Variant [of stringy] and Experiment

Any unification candidate, which can be said to be a variant [of stringy] requires the grand unification scheme or it will yield an endless amount of variant [of stringy]'s. [1] Any unification candidate that is not analogues to a variant [of stringy] may comply with the EOC Guideline but is not subject to the grand unification scheme [2]. Rather it hold no plausibility to experimental stringy.

References

[1] Sanchez-Rey, Miguel A. The Physicalist Program. Createspace: 2015.

[2] Sanchez-Rey, Miguel A. Suitablity and Non-Suitability of Certain Unification Schemes for the Grand Unification Scheme. Vixra.org: 2016.

An Interplay and the Scientific Method

By

Miguel A. Sanchez-Rey

The building blocks of matter and force particles are said to be one-dimensional strings

that reside in a compactified 11-dimensional Kahler manifold [1]. When these strings tangle,

twist, and twirled with each other different matter particles are constructed while their unique

vibrations signify any possible of the five known forces of nature [electromagnetism, weak,

strong, nuclear, and gravitation] which are particles with zero mass giving spin-2 properties for

graviton and spin-1 for all the other force particles [2].

Constructivism is very much a key element of the physical sciences. Certain molecules

are constructed by the electrical attraction of atoms. The D.N.A. strand building blocks are base

on three unique nucleic acids. Cells are manifestations of genes that make up X or Y

chromosomes [3].

In the engineering and experimental sciences certain tasks are base on using principles

of mechanics and dynamics to achieve technological advances. As well certain experimental

advances can also lead to theoretical advances.

In the ever increasingly complex digital era, with advances in space science, bio-

engineering, material science, and etc., the scientific method is still very much a productive

methodology. But the scientific method is a tool for experimental acquisition. As well

constructivism is very much a natural part of the technological sciences but it's limitations are

not without notice. Not all things are imaginably creative.

We can attempt to construct artificial intelligence but we have yet to learn, using the knowledge of today, the cellular rules that govern the brain modules and interface [4]. We can attempt to construct warp drives but we have yet to understand more about how to harness warp bubbles and whether warp technology can withstand the accumulation of heat that can melt a star-ship or the gravitational fields that can tear the star-ship apart [5]. We can construct flying cars but the aerodynamics are subject to wind dynamics that make it much more difficult to achieve.

Ambitious tasks in the technological sciences are presently a slow endeavor and require the cooperation of hundreds of scientists, applied mathematicians, engineers, technologists, and tech industries and academic research institutions. Funding can also impose limits.

The exchange between constructivism and the perspectivism can be said; then, to be an interplay. An interplay in which the experimental, mathematical and theoretical sciences jointly pursue a set of giving problems that results in full integration; whether it's, currently, to answer questions about D-energy, super-symmetry, and meta-space [6].

And then there is the joint venture of the technological and mathematical sciences that tries to pursue questions about measurement, modeling, computer simulation, and applications. Using all we can devise the applied mathematical sciences.

The theoretical sciences are an interplay between physical principles and mathematics. Theoretical mathematics is an interplay between axiomatic set theory and topology. We can then derive mathematical physics.

It's then that the specialized sciences can work together to achieve advances in technology and theory. To answer questions about the nature of the universe and to speculate [in the realm of philosophy] the epistemology of existence and truth. The edge of human knowledge is answerable if an interplay is imposed in the investigation of philosophical understanding.

An interplay doesn't have to be subject to just constructivism and perspectivism. It can also be the exchanges of reductivism, emergence, complex and information systems.

No one dares deny the dominance of logical and methodological thought but change in strategy is needed if changes to perspectives in scientific and technological reality is needed to make feats in the engineering sciences and to answer important questions about the quantum universe and or even the phenomenology of human perception [7].

Interplay is a novel type of gaining control of a rapidly increasingly complex information system. Where hundreds or even millions [both computationally and biologically] of data are being processed and used to ascertain key experimental and theoretical questions. By resorting to interplay one cut's down on the amount of time it takes to answer these questions by using different strategies all at once very much like a master chess player can devise different strategies to checkmate a king or a go player can use different strategies to defeat an opponent by taking all it's go rocks at all once.

The benefits of the strategy of interplay are limitless and powerful. By thinking in terms of strategy and possibility, with the scientific method, one achieves leaps in human and technological thinking where no bounds to imagination are seen but in which wild ideas are tempered by restrictions in physical reality.

References

[1] Deligne, Pierre. Quantum Fields and Strings: A Course for Mathematicians. American Mathematical Society: Institute for Advanced Study: 1999.

[2] Varadarajan, V.S. Supersymmetry for Mathematicians: An Introduction. American Mathematical Society, 2004.

[3] Alcomo, I. Edward. Biology. 1995.

[4] Matthews, Gary G. Neurobiology: Molecules, Cells, and Systems. Blackwell Science, Inc. 2001.

[5] Lobo, Francisco S.N., Visser, Matt. Fundamental Limitations on 'warp drive' spacetime. 2004.

[6] Sanchez-Rey, Miguel A. Meta-space Energy in the Grand Unification Scheme. Vixra.org: 2016.

[7] Dreyfus, Hubert L. What Computers Still Can't Do. MIT Press: 1972.

Relevance of Infrared Divergences to D-energy

Miguel A. Sanchez-Rey

Enter abstract title here

Is infrared divergence irrelevent to D-energy?

June 6th, 2016

If D-energy meta-states E = ▲ + ▼ + ... s.t. W = [   ] + [   ] + [   ] $\Longrightarrow$ W = E ; then, p [n] $\xrightarrow{\text{C2R}}$ M $\Longrightarrow$ M $\in$ $W_{l,l}$ s.t. p [n] $\xrightarrow{\text{C2R}}$ [   ] + [   ] + ... [   ]. Then $\int$[   ] + [   ] + ... [   ] in which from external control to internal control $\int$[   ] + [   ] + ... [   ] $\longrightarrow$ $\int$[   ] + ... [   ]   we impose an infrared cutoff such that $\int$[   ] $\longrightarrow$ [   ] $\Longrightarrow$ internal control so that E = [   ]. [1, 2]

We observe infrared divergence through computational control parameters within internal control but we can impose a cutoff as shown. There we reach internal control, by then, through SUPREME, by C2R, the grand unification scheme [1]. Implying no logical relevance to infrared divergence in metaspace. But we can show control of D-energy entropic metastates by imposing SUPREME.

References

Sanchez-Rey, Miguel A. Current Mathematical Theory in PHPR. Vixra.org: 2015.

Sanchez-Rey, Miguel A. Meta-space Energy in the Grand Unification Scheme. Vixra.org: 2016.

Religious Naturalism

By

Miguel A. Sanchez-Rey

There is never the ignorance that the atheist lives within a cave striving to reach the

light that reveals the form which is the world-of-truth. The Platonic realm that is ever so often

the guiding force that fights against the ignorance of the shadows that permeate at the other

end which is the fire.

The world of today is a complex system with a fragile eco-system and nearly 8 billion of

human inhabitants that walk across planet Earth striving to live their lives, however, their

turmoil and suffering may inhibit the happy life.

Since the dawn of the Enlightenment, in Continental Western Europe, philosophers

were ever more intrigued by the laws of nature and mathematical certainty of Euclidean

geometry and the Cartesian methodology.

Those, then, felt that reason can propel progress and rid the civilized societies of the

tyranny of monarchy and oppression. Set the stage for parliamentary democracy and topple

the depravity of poverty, starvation, illiteracy, and damnation. All human beingd, born with

natural rights, are not subjects but participants, or citizens, of a democratic society.

The project of the Enlightenment set forth the emancipation of human beings from monarchy and unreason. It gave way, at the peak of American realization, independence from the British aristocracy.

How then should the atheist today live in a world ever more religiously attached and secularly suspicious? Atheist then and now are proponents of progress and sciences. They are the architects, as they say, of freedom of speech and the right to live their lives freely as long as they oblige by the standards of government and laws of civil society. But yet, like the emancipation that usher the American independence, they are oppose tyranny and oppression. They are, nevertheless, motivated to spread the atheist world-view of the truth of superstitious and religious fundamentalist ideology.

There are many variants of atheism: Naturalistic Humanism, Secular Humanism, New Atheism, and Militant Atheism, and etc. All very similar to each other besides their tactical mechanisms. Mechanisms that, similarly, on both ends, express the freedom to practice atheism and the right to advocate atheist thought as to delegitimize superstitious non-sense, creationism, and unquestionable ideology.

Today the planet seems rather strange. A planet in which progress, seems, though true, to be making progress but in which progress is meld with fundamentalist ideology and state worship. Various creeds are ever more reluctant, or more fundamentally, hesitant to question their life-world. Their life-world, unique in many ways, is a phenomenological conundrum:

they observe the same reality; or the same epistemology that is foundationalist in nature.

Whether it be the word-of-god, logo-centrism, truth, or sense, they see a world that is

inherently progressive, messianic, frightening, accelerating and chaotic.

The world-of-today and the world-of-tomorrow are contrasting binary oppositions.

How then do they make sense of such vast worldliness? Scientists apply the scientific method

to gather principles and natural laws, mathematicians apply axiomatic set theory to understand

the language of nature, philosophers study the edges of human knowledge and all others live

their lives day by day without any thought of what is out beyond our senses and reason.

Atheist, of today, are fundamentally oppose to such uncertainties which has fed into

superstitious feelings and flame the fire of bitterness and animosity. Some atheist, like New

Atheism, in conjunction with Militant Atheism, have develop the strategy of spreading atheism

through radicalized tactics. Humanist spread atheist thought by using a more diplomatic

approach. And other atheists are scientistic in nature and academically withdrawn from

atheistic activism.

Militant Atheisms founding, in which a crisis emerged within the Atheist community as

to the legitimacy of atheistic thinking, scientific progress, and state secularism, in a world of

intolerance towards atheist and free-thinkers, was a desperate attempt to rely on extremism

and violent opposition against fundamentalism in general. Their intolerance spread like wildfire but their ideology remained ever more questionable to leading and reputable scholars.

They, inevitably, became a threat to the establishment and political reformers.

The extremity, in which New Atheist propelled their devotional belief in social

Darwinism, hid itself in justifiable practices of advocacy and integrity, but in the end, proved to

be a catastrophically genocidal movement that ever more brought in new followers ignorant of

the larger implications of New Atheisms controversial nuisances.

It only dawn that the New Atheists tactical approach was to take advantage of the

economic crisis as though a crisis of conscience in religious fundamentalist thought is the root

behind in which bad-decision making that led to austerity, is, in nature, the bad-decisions of

religious economists and war-worshippers.

New Atheism, ever more progressive and scientistic, became, yet again, a victim of

ideological and extremist fundamentalism.

However how should fundamentalism be interpreted, and is the intolerance of New

Atheisms anarcho-capitalism and social Darwinism a product of such fundamentalist mind-set?

For New Atheist are free-thinkers, proponents of progressivism, in opposition against barbarism

and the Hobbesian state-of-nature [from a superficial sense]. But they are also militants; in

opposition of the diplomatic approach. They are nevertheless, in contradiction, fundamentalist

in practice and in thought.

Religious Naturalism is in, general, a new type of advance atheism. A new type of advance atheism that is phenomenologically restrictive and skeptical of a deity's importance to atheism and theism. That paradox is to ensue, within the natural world, if both sides ideological beliefs is to be, either/or, the dominant paradigm. And yet, even then skepticism toward theistic ideology, today call agnosticism remains more of a skeptical truth. But how so would an agnostic ask? It's only that the thought of god's existence could drive the human being into the depth of ambiguity.

So the resolution between tolerance and bigotry becomes the project of Religious Naturalism in the scientific state. Which is a harmonious planetary society that is ever more quiet and prosperous. Celebratory, as an umbrella theology, of global and natural history. Fundamentally tolerant and respectful of religious differences and diplomatically center. Not giving way to the religious state but strives to be anarcho-syndicalists as to drive out power structures that induce racial and class indifferences and replace them with self-management and the democratic system of workers' councils and trade groups in a federalized and advance industrial society.

So only then the dismantling of the state machine would give way to a new form of religious society. One in which the religious state is no longer operative but only in which the ignorance of a god's existence is universal and in which natural history becomes the substitute. How should a religious naturalist society coincide, and co-exist, with the inevitable outcome of

Anarcho-Syndicalism. Only that god doubts himself as much as wild strength becomes self-determined and incalculable; without racial borders and neither careless in its sexuality or indifference to experimentation and skepticism.

Energy Scale of the Grand Unification Scheme

Miguel A. Sanchez-Rey

Abstract

Why exactly should energy rise as we head further into the grand unification scheme [GRS]?

June 19th, 2016

Energy Scale for GRS

Raising energy, as we head further into GRS, avoids infrared and ultraviolet divergences, allowing charge monopoles to bond. There is a limit to such high-energy scale in which we can calculate to be D-energy:

$E = \blacktriangle + \blacktriangledown + \dots$ s.t. $W = [\ \ ] + [\ \ ] + [\ \ ] + \dots \implies W \equiv 1 + 2 + 1$ by the TrH Theorem $\implies W^2 = 16$ TeV to be the maxima energy limit for the grand unification scheme [1, 2].

We can calculate minima energy limit to be external control to internal control [3]:

external control $\{ [\ \ ] \longrightarrow \int 1 + 2 + 1 \longrightarrow \int 2 + 1 + 1 \longrightarrow \int 1 + 1 \longrightarrow \int 1 \longrightarrow [\ \ ] \}$ internal control.

References

[1] Sanchez-Rey, Miguel A. TrHT in the Grand Unification Scheme. Vixra.org: 2015.

[2] Sanchez-Rey, Miguel A. Meta-space Energy in the Grand Unification Scheme. Vixra.org: 2016.

[3] Sanchez-Rey, Miguel A. Internal and External Control in PHPR. Vixra.org: 2015.